The Quick Guide To Starting a Business

Using FREE and cheap Internet Resources

By Nnamdi G. Osuagwu

The Quick Guide To Starting a Business
Using Free and Cheap Internet Resources

Copyright @ 2009 Nnamdi Godson Osuagwu

All rights reserved.

Front Cover Design by Nathaniel Thomas

Interior Book Design by Carrie Neeson

Copy Edit by Sonia Canzater

ice cream MELTS

Ice Cream Melts, LLC

ISBN: 978-0-9797480-4-2

Published by Ice Cream Melts Publishing

info@IceCreamMelts.com

TABLE OF CONTENTS

"Let's Get It !!!"

— Nnamdi G. Osuagwu

INTRODUCTION ┝━━━━━━━━━━━━━━━━━━━━

L et me start by saying that I'm a fan of the "Entrepreneur." It takes a certain amount of bravery to invest in an unproven vision. The odds are stacked against this venturous pioneer from inception. Passion, at times, is the only motivating factor. Work can transcend to odd hours in the morning, transforming sleep into a rare treat. Success can mean breaking even while ironically, failure can equate to bankruptcy or a total loss of life savings. Faced with such risk, the entrepreneur manages to move forward, full steam ahead, with an unyielding drive.

There are so many obstacles to face when venturing into business ownership. The goal of this book is to show you the most effective, inexpensive and/or free techniques to get your business up and running.

Over the years, I've been asked by friends, family, and clients for advice on starting a business. This book was inspired by their questions. It can also help existing businesses add value to their current operations by taking advantage of **free** resources found on the Internet.

Here are some common mistakes made by start-ups:

■ Using free email services such as Yahoo, Hotmail, Gmail, AOL, NetZero, etc. A business specific email address (i.e. Your_Name@Your_Business_Name.com) presents a professional image and also serves as a marketing tool with each email sent.

- Using a home phone number as both a business and fax number. A simple 10 second fax can take up 30 minutes with busy signals and constant redialing.

- The ability to only receive checks in your personal name as opposed to your business name.

- Not having a method (i.e. website) for a potential client to learn about your services when you are not present.

Trust me, the list goes on and on. I've actually been there, and some of the above examples were my own business. Yes, I will admit to being a novice early on and making critical mistakes that cost me potential clients. The sad thing about such situations is that most potential clients will not tell you why they are not using your business. They may not mention that your business operation is not up to the standard with which they are used to dealing. They will just not call you back and you will never know why.

Having a successful business is not an overnight process. It takes more than this book can give you. Starting your business is only the beginning. There are many obstacles with which to deal while you are in business. This book is meant to give you a good foundation.

Here is a little bit about my background: I am a serial entrepreneur with a background in computer science who has started businesses in the following industries: Information Technology, Publishing, and Real Estate.

To date, I own an international Information Technology Consultancy, a Publishing/Production company, and a Real Estate Development Firm. Throughout the book I will provide case studies and real life examples of my business ventures, and I will cite clients who I have helped along the way.

By the time you reach the end of this book, you will have the tools and instructions necessary to start your own business and present a professional face to the world.

We will cover the following topics:

1. Setting Up Your Business

2. Opening a Business Bank Account

3. Introducing Your Brand

4. Basic Goals of a Website

5. Blogging and Video Blogging

6. Leveraging Online Social Networks

7. Free Tools Offered by Google

8. Free and Cheap Resources to Advertise Your Business

9. Virtual Mailboxes

Having a successful business is not an overnight process. Passion, at times, is the only motivating factor.

"If you WANT IT, then do everything in your power to GET IT. Life is not passive, it's a contact sport."

— *Nnamdi G. Osuagwu*

CHAPTER 1.

SETTING UP YOUR BUSINESS

I t is great that you decided to start your business! There is nothing better than exploring and excelling at your passion. In order to accept checks from clients or customers in your business name, you have to form an entity. There are many types of entities that can be formed. Here are a few:

C Corporation

S Corporation

Limited Liability Company (LLC)

Doing Business As (DBA)

I will not pretend to be a lawyer or an accountant. My goal is to provide a brief overview of the above types of entities and online solutions to get them established.

The definitions that follow are from the following link:

http://www.allbusiness.com

C Corporations

A "C Corporation" is a business organized as a separate legal entity with ownership evidenced by shares of stock. The corporation is formed by filing the articles of incorporation with the state authority, who returns it with a certificate of incorporation; the two documents together become the corporate charter. Each founding shareholder receives from the company a specified number of shares of capital stock. A shareholder may sell owned shares to other investors. The corporation is a legal entity separate from its owners. Advantages of a corporation are the ability to obtain large amounts of financing through a public issuance, ease of transferring shares, limited liability of owners, unlimited life, and professional management.

There is a potential of double taxation with C Corporations. Essentially the entity, being the corporation, is taxed along with profits passed to shareholders as dividends. Accountants have all sorts of techniques to avoid such taxation issues. One common technique is to zero out profits with salaries or bonuses. So if the corporation brought in $100,000, then the salaries and bonuses should equal $100,000 to avoid double taxation. This is just an example, but a qualified accountant will provide a more detailed analysis.

S Corporations

An "S Corporation" is a form of a corporation whose shareholders may be taxed as partners. That is, income is taxed as direct income of the shareholders, regardless of whether it is actually distributed to them. To qualify as an S corporation, a company cannot have more than 35 shareholders; it cannot have more than one class of stock; it cannot have any nonresident foreigners as shareholders; and it must properly elect S Corporation status. The key advantage of this form of organization is that the shareholders receive all the organizational benefit of a corporation while escaping the double taxation of a corporation.

Some consultants with long term contracts get paid

through their S Corporations. Many Information Technology consultants, for example, are given the option of becoming employees or independent subcontractors. The consultants that choose the option of an independent subcontractor have the option of forming a one person S Corporation. The income derived from the newly formed S Corporation is taxed directly to the consultant's personal income as per the definition in the previous section. We will discuss the election form that needs to be filled out in order to convert a Corporation into an S Corporation.

Limited Liability Company

A 'Limited Liability Company (LLC)' is a business form that provides limited personal liability, as a corporation does. Owners, who are called members, can be other corporations. The members run the company unless they hire an outside management group. The LLC can choose whether to be taxed as a regular corporation or as a pass-through to members. Profits and losses can be split among members any way they choose. The LLC rules vary by state.

Shareholders' personal assets are protected in the event of business-related lawsuits. The tax situation for this type of company is much like that of the partnership in that it acts as a pass-through tax entity. A tax return for a partnership is filed with the IRS for information purposes only. All income and expenses are attributed to the shareholders of the LLC. According to the LLC agreement, the shareholders can allocate income and its resultant tax liability in the same way as partners in a partnership."

My real estate development firm, Sensible Developments, LLC, was formed as a Limited Liability Company. My accountant advised us to form an LLC because the taxes can be passed onto our individual personal income. Limited Liability Companies are not taxed directly on an entity level. The concept of pass-through taxation means that taxes are passed down to the owners. Pass-through taxation is used both for Limited Liability Companies and S Corporations.

We chose the Limited Liability option for Sensible Developments, LLC because in addition to the pass-through taxation benefits, personal assets are shielded from business debt in most situations.

Doing Business As, or "DBA"

Doing Business As, or "DBA" is an assumed name a person uses for a business instead of the actual business name or one's personal name. A certificate should be filed in the courthouse to use an assumed name, and to assure that no one else in that jurisdiction is using the same name.

I used a DBA for my consulting company, Strategic Generation. Since I had existing companies, I didn't want to form yet another Corporation or Limited Liability Company. My accountant advised that I could use one of the other companies as a base and form a DBA. I also needed the DBA paperwork in order to open a bank account and accept checks in the business name. Also, note that you can form a DBA using your personal name if you don't have an existing business.

Once you decide on the type of entity that you want to form, here is an online resource to get it done for an inexpensive price:

The Company Corporation (http://www.incorporate. com)— I have used this resource quite a few times to start various business ventures. Their packages range from $99 - $399 for US based businesses. They also handle international corporate packages. For the purpose of this exercise we will focus on domestic packages.

The cheapest way to utilize their service is to get the basic $99 package and file the necessary government forms directly with the IRS. Here are the additional forms that you will need, if you elect to go with the cheapest package:

1. **Subchapter S Election Form 2553**—This form needs to be submitted to the IRS if you choose to form an S Corporation. Form 2553 can be found by going to the following URL:

   http://www.irs.gov/pub/irs-pdf/f2553.pdf

Another option is to "Google" (http://www.google.com) Subchapter S Election Form 2553

2. **Employer Identification Number or Tax ID Number**—It is similar to a Social Security Number but is for your business. Your business is its own entity and can be treated as such by the IRS, hence the tax id, The EIN number is available **free** by going directly to the IRS.

You can apply via the internet:

   https://sa1.www4.irs.gov/modiein/individual/legal-structure.jsp

or via phone:

Taxpayers can obtain an EIN immediately by calling the Business & Specialty Tax Line at (800) 829-4933. The hours of operation are 7:00 a.m. - 10:00 p.m. local time, Monday through Friday. Note: International applicants must call (215) 516-6999.

Please also note the general rule with most business filings is that they are relatively cheap or free if you go directly to the government entity. Understand that although it is cheaper to deal directly with the government, it is also extremely detail-oriented – one mistake and they will ship your items back to you for a redo. However, I do not think that should discourage you from trying. Practice makes perfect, and you will save in the long run by learning the process.

Now, if you have an existing business but want to use a different name for a new line of products or services, then you can form a "DBA" or "Doing Business As." As I stated earlier, my accountant suggested that I form a DBA for my consulting company, Strategic Generation. The original company is still in existence and operating. I wanted to keep the two brands separate, but didn't want the hassle of forming a completely new entity. Here is a site that provides links to each state's DBA filing requirements:

http://www.business.gov/register/business-name/dba.html

It is extremely important to reiterate that details are extremely important. It cost me $25 file the DBA for Strategic Generation in New York State, but they kept sending it back as incomplete because I forgot a line or forgot to check off an appropriate box. I ended up spending an additional $25 for expedited service, which totaled $50 to file Strategic Generation as a DBA. It was still a lot cheaper than having a company or lawyer do it for me. Some online agencies quoted $99 plus the state filing fees. So I ended up saving at least $99.

HINT: Check the availability of your business' domain name before registering the entity (Go to http://www.godaddy.com to check availability). It is your choice to file if the name is unavailable. I usually try to file company names that have available URLs. The domain name is the address used so that people can find you on the Internet. We will discuss this further in the next chapter. Most for-profit businesses have a .com extension and non-profits have .org.

Examples: IceCreamMelts.com
StrategicGeneration.com
SensibleDevelopments.com

CHAPTER 2

OPENING A BUSINESS BANK ACCOUNT

Terrific!! You have now filed the necessary paperwork to start your business. Congratulations. The next step is opening a business bank account and gearing up to accept payments.

A bank account serves a lot of other purposes besides payment acceptance. Establishing a business banking history can help for future lines of credit and loans. Money is like fuel when it comes to your business, so the more fuel your car has the further you will be able to travel.

When you open a business bank account, the bank representative will ask for the following:

1. *Articles of Incorporation or Articles of Organization (LLCs)*

2. EIN Proof – sometimes just the number will suffice; it depends on the bank with which you choose to work.

3. Filing Receipt

4. Personal ID – they will usually make a photo copy of your driver's license.

The bank will provide a list of required documents. All of the documents collected during the formation of your business should be adequate.

It can become cumbersome to keep all of your business documents in order. An easy way to keep track of all of these documents is to convert them to electronic files. This can be accomplished in two ways. The first option is to scan the documents and save them as PDF files. The second option is to fax the documents to your e-fax number and save the fax as a PDF file. *(We will cover e-faxes in Chapter 3. Introducing Your Brand).* Once the documents are all in electronic form, another trick is to save them on a portable flash drive or email them to your smartphone. Portable flash drives and smartphones will be revisited in *Chapter 10. Conclusion.*

The next step is to decide on the bank or banks with which you will do business. Business is about relationships; based on that concept, I personally believe in opening bank accounts in as many banks as possible in order to establish new relationships – well, at least the ones that don't charge annoying monthly maintenance fees. Establishing banking relationships has helped me tremendously. Every bank is different, and some excel in different aspects of their business. I've been denied credit at some banks when other banks gave me credit. Although the outcomes were different, each bank representative asked the same question before starting the loan application, "Do you currently bank with us?" Each time I answered, "Yes," and provided my account number. Bottom line, big or small businesses treat current customers more favorably than new customers, especially when it comes to making an exception. People tend to be strict with rules when they are dealing with people with whom a relationship has not been established. People and businesses are the same in this respect.

No matter your decision, it will be prudent to review their website and make sure their banking terms are aligned with your business. Another important factor is for you to get to know your branch manager.

I never understood this concept until I met an older gentleman by the name of Mr. Reddick. He was previously a branch manager for a trusted bank but has long since retired. He did business in a time when a handshake, your word, and a smile could close a mortgage deal. Those times are long gone, but the lesson is not. He said that branch managers have the power to make exceptions in certain cases when dealing with clients, but they, of course, will not make exceptions for people they don't know. I'm not saying that you need to invite your branch manager to dinner (although I did invite one to a book signing), but at least schedule an appointment to meet and discuss your business needs. *The manager will be delighted to meet with you and if not, for some strange reason, then you should think twice about doing business with that institution.*

No matter your decision, it will be prudent to review the bank's website and make sure their terms are aligned with your business.

I have several bank accounts and established relationships with several branch managers, but one of the most memorable branch managers that I have ever met was at a local bank. I was treated like a rock star when I opened my business account for Sensible Developments, LLC.

We had prior contact because I called ahead. She took the liberty of visiting our company's website and read through our bios, company mission, and past projects. When I arrived, she set up a meeting with the branch's business banker. In the meeting, the branch manager asked me in depth questions about my business. The questions were far from generic; I immediately noticed that she did her homework on Sensible Developments.

I was totally impressed and delighted to open an account with her branch. Their nonexistent monthly fee for business accounts was just a bonus. I must also add that they were both impressed when I emailed them right from my phone all of the documents that they required. We will further explore this concept in *Chapter 10. Conclusion.*

Since that meeting, I have opened two additional business accounts in that branch and also a personal account. What is even better is that I can go directly to the branch manager. It reminds me of my favorite scene in *Goodfellas* when Ray Liotta's character, Henry Hill, walked through the kitchen of a crowded restaurant and the manager set up an impromptu table for him and his date. I'm not saying that you will get the same treatment by knowing your branch manager, but you are definitely one step closer.

Let's now talk about monthly fees. Read the fine print and ask questions. I have one basic rule in business that took me forever to learn:

HUGE RULE

Get details on everything that you do, and if you don't understand something, just STOP. Proceed only when you are fully on top of your game and have a full understanding of the venture. This especially applies to banks. This is a rule that I sometimes forget to implement due to impatience. It is a work in progress for me.

It is very important to watch out for monthly fees. I've gotten caught so many times with different institutions. Now, if you are depositing over a certain amount, then fees can be avoided, but if you are depositing under $50 on a shoestring budget, then it is important not to get charged a $25 monthly fee.

PayPal (http://www.paypal.com) – Once your bank account is setup, having a PayPal account can be a useful tool for accepting credit card and online payments. PayPal is an online e-commerce business that allows money to be transferred between entities (i.e. individuals or businesses). After their verification process, you can connect your business bank account to your PayPal account and use it to accept credit card payments online. PayPal charges approximately 1.9% – 2.9% + $0.30 USD per transaction. You only pay them when you use their service as opposed to incurring monthly charges with a merchant account. I use their services to pay my vendors, accept payments from clients of Strategic Generation, and collect payments from Ice Cream Melts customers that purchased books and t-shirts.

So the important notes from this chapter are the following:

1. Read your bank's policies on their website.
2. Decide which branch you will use to open your business account.
3. Call ahead and schedule a meeting with the branch manager.
4. Ask questions in the meeting until you fully understand their policies.
5. Bring all of your business documents, whether digital or hardcopy.
6. Establish relationships with as many institutions as possible.

> **"Confidence, balanced with humility, equals a powerful form of sincerity."**
>
> — *Nnamdi G. Osuagwu*

CHAPTER 3

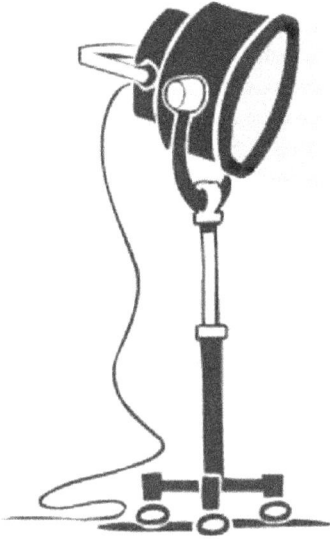

**INTRODUCING
YOUR
BRAND**

Y ou now have an established business and a business account in one or more banks with branch manager relationships in each. Wonderful! Now it's time to network and promote your business. The first part of any marketing campaign starts with you, the owner of the business.

This is your passion. If you don't promote it, who will? For the purpose of this book, we will focus on individual branding and marketing efforts.

Let's start with a basic business networking event. You just had a conversation with a potentially great contact. The goal is to follow up and keep in touch. At the end of the conversation, the contact gives you a business card and then you give yours. For some self start-ups, this presents a slight problem. Here are some of the potential issues:

1. You don't have a business card and must write your contact information down on a piece of paper or enter it in your phone.

2. Your business card lacks a consistent design across your marketing material (i.e. brochures, flyers, website, etc.).

3. Your business card may just have your name and cell phone number.

Remember, no matter how small your business may be, the goal is to look like an established Fortune 500 business. Let's solve these issues by doing the following:

1. Decide on a logo.

2. Set up a virtual answering service for phone and fax messages.

3. Choose a domain (URL) for your business.

4. Decide on a basic business card layout.

5. Get your business cards printed and ready for distribution.

Deciding on a Logo

Logos are extremely important. It is a symbol that represents your company. Some of my favorite and recognizable logos from successful businesses are:

1. **Apple, Inc.** — I think the bitten apple was brilliant and could symbolize so many different things.

http://www.apple.com

2. **Citibank** — I like the way the red bow connects both the "i's" in "citi." The "t" in the logo is being protected by the bow. Multiple meanings and symbolisms can be drawn from their logo. It is also memorable.

http://www.citibank.com

3. **Temple University** — I'm totally biased because this is my alma mater, but I always loved the way their logo stood alone on a sweatshirt. It is easily recognizable.

http://www.temple.edu

The above logos can be viewed on the Google search engine (http://www.Google.com). Type "Business_Name logo picture" (i.e. Citibank logo picture) to see actual pictures of the logos listed above. Most logos from big companies are trademarked and require explicit reprint permission.

Here are some of the logos from the companies that I have started:

Ice Cream Melts, LLC:

ice cream **ELTS**

Our goal was for the upside down ice cream cones to stand alone and still symbolize the brand. The cones resemble an "M," and they are also melting, which is part of the last word in the company name, Melts.

Strategic Generation:

Strategic Generation

This logo is a little more conservative. We decided to spell out the complete company name but in addition have a stand-alone graphic that could represent the company.

Stand-alone logo images can also be used to create favicon files. These files usually have an .ico extension. They are the small graphics that you see on the tabs of your browsers when accessing a website. The Internet Explorer browser does not seem to present favicon files, but if you use Google Chrome, Safari, or Mozilla Firefox and open http://www.IceCreamMelts.com you will be able to see our logo on the tab of your browser.

Here are some online resources where you can get a logo for a reasonable price:

1. **Logoworks (http://www.logoworks.com)** – their basic silver package is their least expensive package starting at $299.00. I have used them prior to starting Strategic Generation. Now we do all of our own logos in-house through our consulting firm.

2. **Strategic Generation (http://www.strategic generation.com)** – A major reason for starting Strategic Generation was to provide entrepreneurs with end-to-end business solutions at reasonable prices. Our basic logo service starts at $250.

Setting up a virtual answering service for phone and fax messages

Most established businesses have both a phone and fax phone number. My goal is to make your shadow appear larger than your actual size. So let's start with the business phone number.

It is important to be able to receive both business voicemails and faxes without interruption. Here are some services that I have used for my business ventures.

Accessline (http://www.accessline.biz) – This company has many phone related services. The one service that I have used for about 3 business ventures is their voicemail and fax service for $8.95 a month with applicable taxes. It is really straightforward. You receive the same local number for both your business voicemail and fax. Someone can call or fax concurrently. You will then get an email or text notice regarding your message or fax. Once you check your message, you can then get back to the client via your cell phone or landline.

It is quite common for people in offices to let incoming calls go to voicemail. Why does anyone have to know that you didn't temporarily step away from your business land line? My favorite voicemail is from our Real Estate Development business, Sensible Developments, LLC. Call and listen to the message: 215.525.1052. The operator is clear and presents a corporate image. This is a typical Accessline number.

An old colleague got an Australian friend to leave the message. I think that I will keep that message as long as the business is in existence.

Did I mention that there are no hidden charges with this company? The rate is a fixed cost once you apply the applicable taxes. Fixed cost line items are great because it helps us independent business owners do the magic "B" word. Yes, BUDGET.

Tip: You can sign up for the service through their website, but if you call and speak to a representative, he/she may be able to waive the initial registration cost, which is around $25.

Kall8 (http://www.kall8.com) – Another company that has many telecommunications features. I only used their $2 per month toll free number. Yes folks, I said $2 per month.

Ok, here is the skinny on this company. Conceptually, they are the same as Accessline in terms of how the number aspect works. The only difference is that this service is not fixed. You are charged when people call you and leave a message. The only fixed cost is the $2. Everything else is variable.

Having a toll free number definitely sets you apart. It is not necessary for a startup but great if you can afford it for your venture. I don't have a toll free number anymore because it was an unnecessary expense and my clients didn't seem to mind calling the local number.

A fellow small business owner/art gallery friend of mine has it and loves the service. She actually referred me to it. Her business deals a lot with exclusivity and maintaining a certain high-end look so the toll free number adds to her presentation. You have to decide if it will add to your presentation.

Choosing a domain (URL) for your business

This is the fun part. As mentioned in the previous chapter, you should have checked the availability of your business domain prior to registering the entity. Go ahead and buy the domain if you have not done so already. I use the following service for purchasing domains:

Godaddy (http://www.godaddy.com)—Domains at Godaddy cost $10.69 when you purchase them through their website. If you have the time and a bit of patience, call and speak to a representative and possibly save yourself $1 or so. Sometimes they have specials that are available on the phone but not on the website. I know it seems small, but every penny counts.

Always remember that names are important and should leave some sort of impression. I remember a classic example. I was at a Mashable (http://www.Mashable.com) networking event. They were voted one of the top 25 blogs in 2009 by Time. I met their COO, Adam, and he absolutely loved my business and domain names. He mentioned that they were great and extremely creative. My names stuck in his mind. We are now following each other on Twitter. How cool is that?

When we talked, I told him about the following domain names:

GlobalBoredom.com — internal application that we built to help people find activities in their local areas.

IceCreamMelts.com — interactive publishing/ production company and the name of my first book.

StrategicGeneration.com — Business and Information Technology consulting company.

Deciding on a basic business card layout

If you have graphic design skills or are comfortable with software such as Microsoft Publisher or Adobe Indesign, then you can essentially layout the business card yourself for **free**. You would insert your logo and choose from their many templates. You should also try to balance all of the colors on the card so that they blend with your theme.

It is important to note that your business card should have the following essential information:

Business Name (i.e. Ice Cream Melts, LLC)

Business Website Address — URL (i.e. http://www.IceCreamMelts.com)

Business Phone Number (i.e. 215.525.1052)

Business Fax Number (i.e. 215.525.1052)

Business Logo (could be same as the Business Name)

Your Cell Phone Number (totally optional)

I am also a fan of making double-sided business cards so that no matter how my card is thrown on a table something is always showing. It also helps differentiate between my card and a bunch of random cards with plain white backgrounds.

On the next page is an example of the Strategic Generation business card, both front and back:

Strategic Generation Front Business Card View:

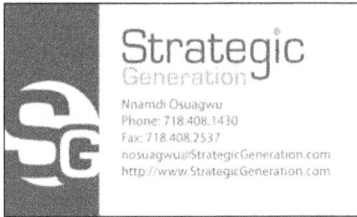

Strategic Generation Back Business Card View:

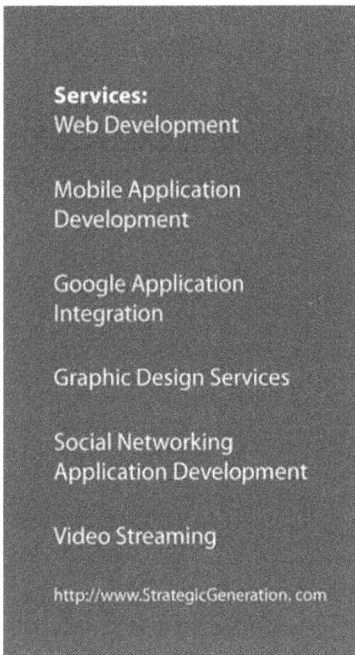

Now, if you just want to get it done and not worry about such things, then here are some online resources:

1. **Market Splash
(http://www.marketsplash.com)**
There is a customized $99.00 option which gives you the 2 original business card concepts and 2 business card revisions.

2. **Strategic Generation
(http://www.strategicgeneration.com)**
We offer 1 business card design and 2 revisions for $75.

Getting your business cards printed and ready for distribution

Now that you have your PDF business card file, you are ready to GO. One of the best resources that I found for printing is PS Print. They are extremely reasonable in terms of price and have superior print quality services. I like them so much that I became an affiliate of their services so that I can get credit when my clients use them. Since their printing prices are low, I am able to offer my clients packages that combine printing and graphic design services.

You can access their URL via our website:

http://www.StrategicGeneration.com/partners

They have specials all the time, but the sweet spot is 1,000 business cards. You seem to get more for your money with the 1,000 card option. Definitely compare prices with all of the options, and see which one works for your budget. Regardless, you should be well under $100 and have enough business cards to last for a while.

"Some think local and some think international ... the power of the Internet allows ANYONE to gain international recognition ... just need the right product, service or message."

— *Nnamdi G. Osuagwu*

CHAPTER 4

BASIC GOALS OF A WEBSITE

We are now off and running. I love it. You now have bank account(s), banking relationships, business cards, phone numbers, and a reserved domain name for your future website.

At this point, you may have the following question: Why print my website URL on my business cards without having a website up yet? What if someone goes there and sees nothing?

My answer to the question is that your website URL is not going to change so it is not a problem. It is better to have everything offline in place so that when you do launch your website, there is no need to then wait for business cards.

It is quite common for new business websites to have a basic "Coming Soon" or the default page given by the company from which you purchased your domain. This is called domain parking. Essentially, you own the domain, and you are keeping it on the domain company's server. In the example of using Godaddy.com, Godaddy is holding your domain on their servers free of charge. They actually benefit by putting advertisements on your domain's page. When I was about 7 years old, I tried to use a coupon in a Brooklyn bodega (corner store). The gentleman behind the register did not accept the coupon and quickly informed me that nothing is free in America. So in the case of domain parking, a company like Godaddy is making revenue from advertisements on your site, and of course, not sharing those profits with you. That is their price for temporarily hosting your domain.

So here are our goals:

1. Find a company to host your site.

2. Post a temporary "Coming Soon" page while your site gets developed.

3. Create business requirements for your site.

4. Get your website developed

Finding a Hosting Company

There are plenty of hosting companies around. If you Google "Web Hosting" you will see a long list of companies. The following site gives unbiased reviews: http://www. thetop10bestwebhosting.com.

The top two sites (as of June 2009) are www.JustHost.com and www.FatCow.com. Both seem to give a lot of freebies, which is always a good thing. Choosing the right hosting company can be confusing because of the sheer amount from which to choose. It basically all depends on your needs.

If your site is strictly for presentation purposes and you are not looking to achieve Facebook or Twitter-like traffic, then one of the cheap shared hosting plans may be your best bet. A shared hosting plan is simply when you are one of many websites sharing space on a server. Some of these plans also offer free domains, meaning you can save the annual domain registration charge as long as you host with them. You will also be presented with Linux versus Windows operating systems. I prefer Linux because most times it is cheaper, and in terms of development, I have access to more LAMP-based developers than WISA developers. LAMP stands for Linux (operating system), Apache (web server), MySQL (database), and PHP or Python (scripting languages). WISA stands for Windows (operating system), Internet Information Services (web server), Microsoft SQL Server (database), and ASP (scripting language).

More info on LAMP and WISA can be found at:

http://www.shawnolson.net/a/302/a-comparison-of-two-major-dynamic-web-platforms-lamp-vs-wisa.html

For my business ventures, I used www.Godaddy.com's Virtual Dedicated Server plan:

(http://www.godaddy.com/gdshop/hosting/virtual.asp?ci=90 13&display=virtual)

and Scorpio Informatics' (https://scorpioinformatics.com) hosting plan for my clients.

A dedicated hosting plan is not shared with other websites.

My reason for choosing www.Godaddy.com's Virtual Dedicated Server is because I wanted the cheapest way to be isolated from everyone else and needed control over my server. Companies that offer shared hosting plans usually prevent installation of software that may affect other users. We also build many internal web projects and wanted all of our projects on one server. If you are building a single site for your business and don't require software installation on the server, then a Shared Hosting Plan is a better and cheaper option.

I use Scorpio Informatics hosting for my clients because they are reliable, and I had a direct conversation with the owner. We offer a maintenance option to our clients, and I wanted to make sure that I had direct access to senior staff in case of any disruptions. Currently, we set our clients up with a $110 annual hosting charge as opposed to monthly payments.

Put Up A Temporary "Coming Soon" Page
While Your Site is Being Developed

Once you've decided on the hosting company and purchased a plan, the next step is putting up a quick "Coming Soon" page. It will take some time for your site to be developed, but in the interim you should have your basic contact information on the site.

Remember, your business cards should be arriving soon, and you will start networking. People will totally understand a "Coming Soon" page with contact information listed. You can even include a short description of your company. I will show you how, or of course, you can have the people developing your site add one in.

The first file read in a static website is an index.html file that is located in the "public_html" folder. So when you type in your URL: "http://www.Your_Company.com," the browser will read and execute the code from index.html file. Our goal is to create an index.html file and place it on your server. I will show you how to code the index.html file and also give you links to a free FTP program. FTP stands for File Transfer Protocol and is basically the way to get files located on your computer copied to a remote server.

1. Copy the following code and place it in a new untitled Notepad (Window Users) or TextEdit (Mac Users) file:

<html>

<head>

<TITLE>INSERT_COMPANY_NAME</TITLE>

</head>

<body bgcolor="White">

INSERT_COMPANY_NAME

INSERT_MESSAGE_ABOUT_YOUR_COMPANY

Phone:INSERT_BUSINESS_PHONE_NUMBER

 Fax: INSERT_BUSINESS_FAX_NUMBER

Full Site Coming Soon ....

</body>

 </html>

2. Customize the above information in the text file and Save As an HTML file:

The file must be saved as an HTML file. By default it will be saved as a text file.

You must select All Files from the Save As Type drop down menu.

Change the name in the File Name box input to index.html

For more information on creating HTML pages go to http://www.htmlcodetutorial.com.

3. Transfer the file to the public_html folder on your server using FTP

In order to use FTP, you first have to download a free FTP program. There are many free FTP programs, but I use FileZilla. Filezilla can be downloaded from

http://filezilla-project.org

Once you download and install Filezilla, you must get

access to your FTP server. You must find out the login details. It is pretty straight forward. Please note the following:

The Host field must be in the form of sftp://DOMAIN_NAME (i.e. sftp://IceCreamMelts.com)

The Port field can be left blank. Filezilla will insert the correct port number (22) when the connection is established. Your hosting company or web developer should be able to provide you with your FTP username and password if you don't have it.

I can also understand if some of you may not want to go through the above steps. Of course, doing such things yourself is cheaper, and it is always good to learn as much as you can about your website because websites and technology are so critical for every business. It can also get costly to have a consultant make every minor change to your site.

Also, note that Information Technology consultants **don't** have an infinite knowledge base or know every technical thing end to end. We are, however, trained to find and learn the necessary information needed to complete any technical task. Such information can be found in books or the internet. I usually Google issues and then look on forums to see if anyone posted solutions.

Most issues are not new and plague others as well. The trick is finding those solutions and implementing them on your end. I think that I might have just betrayed the secret covenant of Information Technology consultants everywhere.

Come Up With Business Requirements For Your Website

I'm not a fan of building a website just to have one. People tend to do this often. I agree that having a site is essential for your business, but I think that at the end of the day, it is technology and should be used to make life easier. Based on that thought, a website should be used to assist in automating your offline activities.

Here is an example of what we did for our construction company, Sensible Developments, LLC. We are licensed general contractors, so we wanted a way to give status updates to our clients. The current process was that we would do the work and then call our clients to give them an update, or the client would call us and ask for an update. This can obviously be time consuming if you are working with multiple clients. So we added a feature on our website that allowed us to post pictures and input notes on our projects. Each client is given a login ID and can view the status of the particular job.

That allowed us to communicate with our clients asynchronously, meaning that I can upload pictures and input comments, while still in bed if I want, and my client can check on the status of the construction project anytime afterwards. Of course, clients can still call, but this action limited the amount of calls made and time spent on each call. This component is a perfect example of how our construction company's website helped automate an offline activity.

Each business is different, but time should still be spent thinking of ways that the website can make your day-to-day activities easier.

Each business is different, but time should still be spent thinking of ways that the website can make your day-to-day activities easier. You can start by making a list of some of your daily routines and common questions from customers or clients. Also, create of list of things that consume much of your time. Prioritize this list, and create business requirements around them. If you have staff, it will also be helpful to include them in this discussion.

The business requirement phase in Software Development is a job within itself that is equipped with professionals such as business analysts who get paid good money to generate requirements and coordinate stakeholder meetings. Stakeholders are usually the people who will benefit from the end product. In your case, this process is not as involved, but should still be taken seriously. It is also helpful to adopt similar principles used by the big boys to generate requirements. A good online article about business requirements can be found at:

http://ezinearticles.com/?Tips-for-Writing-Excellent-Business-Requirements&id=172406

Get Your Website Developed

Now it is time to build your company's website. There are a ton of free applications, but they may not be viable solutions for accomplishing your outlined business requirements. I recommend working with a technology consulting company or web developer to get your site built correctly and in accordance with your business requirements.

Here are some things that you should ask your consultant:

What type of content management system will my site use?

A content management system is important because it allows you to easily update dynamic content behind the scenes. You should make sure that your consultant uses one of the commercially available applications like Joomla (http://www. joomla.org) or Drupal (http://drupal.org). Some consultants opt to build their own home grown content management system. I totally disagree with this approach for the following reasons:

1. The above mentioned content management systems are free. Why build something from scratch if available tools exist?

2. If for some reason you and your consultant part ways, there are a plethora of other consultants with expertise in either Joomla or Drupal who can pick up where the

other consultant left off. Unfortunately, if the consultant leaves you with a home grown content management system, the new consultant may have to start from scratch to expand your site. It is possible for the new consultant to learn the inner workings of your site and still expand, but regardless of the option, both will take more time. In the world of consulting, time equals your money.

We had a client that approached us to expand her site. It was a well designed site, but she wanted to incorporate some social networking features and add a lot more components to her existing e-commerce site. Unfortunately, the previous consultant built a home grown content management system. We would essentially have to port her existing site to a commercial content management system, recreate her existing functionality, and then add new functionality. The cost to do those services exceeded her budget so we could not accept the task. From her stand point, she already spent a pretty penny to build the original site and could not understand why we would essentially have to start from scratch. From our standpoint, we had to charge her appropriately for the time that it would have taken us to complete the above actions. If her previous developer had used a commercial application we could have easily added the new functionality for a cheaper price. Everybody would have won.

Will you manage the site once it is built?

This is important if you are a novice. You don't want a situation where the person builds the site and then disappears. Make sure that the terms of post-production support are defined.

Where are the contracts?

Both you and your consultant should enter into a contract that specifies the work that needs to be done. This is important so each party understands what each is getting into.

It is also important to get all usernames and passwords to your site once it is built. This point should be emphasized because some clients trusted their developer to work on the site end to end. The developer purchased the domain, hosting plan and even built the site. Then some disagreement occurred and the developer disappeared. It is important that you maintain ownership of your domain and hosting plan and provide the developer access. If you are confused, have your developer or hosting company walk you through the process.

CHAPTER 5

BLOGGING AND VIDEO BLOGGING

We went from a hot cup of steam to a tangible business entity: business cards, website, bank accounts, and a few banking relationships. This is excellent progress and you should be proud of yourself. I want to now introduce you to the concept of using Blogs and Video Blogs as viral marketing tools to promote your business.

A blog is a website or component on a website that allows users to enter text, graphics or video as often as they like. The word "blog" can also be used as a verb in which a person is entering the above-mentioned information into the web component.

Video blogs, or "vlogs," are the same concept, but instead of text, a person uploads videos of oneself onto a website on a regular basis.

Some popular blogs get millions of page views a day and make money from advertising or donations. The goal for your business is to use this tool as a free medium to show your expertise in your business area. Sincerity is a powerful tool in business. Customers will gravitate towards people who are sincere and passionate about their business and area of expertise. By voicing your opinion and giving tips on blogs and video blogs, you show that you have a genuine love for what you do. It is not just a job or a way to make money. Take for example this book; it should be apparent that I'm a fan of the entrepreneur and will do what I can to help new entrepreneurs along the way. This is the same idea behind blogging and vlogging for your business.

You can, of course, get your web developer to add a blog or video blog component to your website. This is the optimal solution because users can see all of your information at one location. Another option is to use **free** tools available on the internet to tell your story and lend your expertise. You can then take the link from the free tool and add it to your website. Just make sure that the link opens up in a new window so customers are not completely taken away from your website. The web developer can add this link.

Here are some **free** Blog and Vlog Online tools:

Blogger (http://www.blogger.com) – This is a straight forward blog tool that allows users to create blogs. You can choose whatever name you like as long as it is available. The goal is choose a name that closely matches your business.

For example we used the following Blogger URL for Sensible Developments, LLC:

http://sensibledevelopments.blogspot.com/

YouTube (http://www.youtube.com) – This is a video sharing platform. Most people have heard of YouTube and have watched the countless funny videos on YouTube, but most people don't use it for their business. You can create a channel, which is your user account and then you can add videos. Although we have videos on www.IceCreamMelts. com, we still use our YouTube channel to promote our brand awareness. We place banners on each video to drive traffic back to our main site. It is a good practice to choose user names that are the same or as close to your business name as possible.

We used the following YouTube channel for
Ice Cream Melts, LLC:

http://www.youtube.com/IceCreamMelts

Some of our Ice Cream Melts videos have ended up on websites in Germany, France, China, etc. We found this out later by typing our name in Google. Due to the viral nature of the internet, once you post content to such sites, anyone can grab it and repost. It is important to always superimpose your brand on all of your content in the form of banners so that no matter where it ends up, people see your company's name.

Here are some **free** online tools that allow you to share your expertise:

ehow (http://www.ehow.com) – This is a creative tool that allows people like you and me to contribute 'How To' articles. Basically people write articles like:

How To Tan Smart, How To Increase Your Self-Esteem, How to Replace a Car Battery, etc. Anything and everything you can think of is on the site. This is a great tool for small businesses because it positions you to share your expertise with potential customers. The article will of course have a link back to your site. It will also show up when people Google you or the subject matter.

I will even take my own advice and soon post "How To" articles about topics covered in this book with a link back to my consulting company.

Squidoo (http://www.squidoo.com) — Squidoo is an innovative concept that allows users to gather whatever information they have on a particular topic and create pages called "lenses" for all to view.

Both Squidoo and eHow offer compensation if your article experiences heavy traffic. Although minimal, the avenue for potential revenue is still present.

The whole idea behind these types of tools is to share your expertise and raise awareness of your online brand. The ability to provide such information 20 years ago was only reserved for authors or people with access to reporters, magazine publishers, or other media outlets, but with the online tools discussed throughout this book, anyone can provide useful information to the public. Search engines like Google enable users to find information around topics of interest. So if someone is searching for information about a topic that you posted, that person just may find your article. That person could then pass this article to friends. The friends can potentially pass it to their friends. Your online brand and information are now being passed around like a virus. The impression of you as a Thought Leader in your field is becoming contagious!

Lastly, always provide your website's URL on all content that you contribute so that people can link back to your business.

CHAPTER 6

LEVERAGING ONLINE SOCIAL NETWORKS

A good friend, Aliya, who is a publicist, gave me great advice during the promotional phase of my second book, *A Souvenir for My Mom*. She said you have to build a buzz from the people you know and then expand outwards. My original approach was to start with strangers and have my product circle back to people who knew me, but she was right. Starting with people I knew made more sense. Most times people buy "**YOU**." They don't necessarily buy your product or service. They simply buy "**YOU**."

My mom hired many carpenters over the years to tackle various maintenance repairs to our house in Brooklyn, New York. She only did business with people who were referred by her friends. She never once cracked open the Yellow Pages. She paid for the relationship that the contractor had with someone that she knew. She didn't buy services from a contractor named "Joseph." She bought Joseph's relationship with her friend.

It is now time to bring this chain of relationships back to technology. Social Networking websites give you the ability to automate the above chain of referrals and leverage your network of relationships over the Internet.

The following, from the Journal Of Computer-Mediated Communication, is a definition of a social networking website: (http://jcmc.indiana.edu/vol13/issue1/boyd.ellison.html)

We define social network sites as web-based services that allow individuals to:

(1) constructapublicorsemi-publicprofilewithinabounded system

(2) articulate to a list of other users with whom they share a connection

(3) view and traverse their list of connections and those made by others within the system. The nature and nomenclature of these connections may vary from site to site.

Essentially you can create a pool of people whom you know. At that point you can then see people whom they know and so on. You can then request to be friends with people whom they know. In addition, you can post messages about things going on in your life, business, etc. Everyone is now aware of what is going on in your life. Friends from junior high school, high school, and college purchased my books because they noticed that I had mentioned it on my social networking profile. They didn't buy it because they saw a poster on a train; they purchased it because a friend wrote a book.

Examples of social networking sites include Facebook, Bebo, MySpace, and Twitter. I will describe some of the ways Facebook and Twitter helped me in my various business ventures. Although I have accounts with Bebo and MySpace, I primarily use Facebook and Twitter currently.

Facebook (http://www.Facebook.com) — is a social networking site that enables users to create a profile and add friends from school, work or other areas of their lives. The beauty of the profile creation process is that you are asked questions regarding employment, education, and affiliations. These questions are strategic. Once you have answered all of the questions, the Facebook application then suggests people as "friends" whom you might know based on the years that you attended certain schools or were employed at certain places. Before you know it, you are chatting with a childhood friend from elementary school.

Social Networking websites give you the ability to automate the chain of referrals and leverage your network of relationships over the Internet

Besides the aspect of reconnecting with old school chums, this application is a powerful word of mouth marketing tool. Think about it. Let's say that you just started your "Widget" business and want to let everyone know about it. You can easily post information about your business on your profile. Facebook even allows you to create "Groups" and "Fanpages" associated with your business.

Here are some examples of things that I did in the past to promote to various business ventures to my Facebook network:

Facebook Group Example:

My second book, A Souvenir for My Mom: First Hand Accounts from the 2009 US Presidential Inauguration, debuted around Mother's Day of 2009. The book's retail price was $9.95. My marketing budget was pretty slim, and I was racking my brain for innovative ways to get the word out. I decided to create a Facebook Group titled: Let's Create the Longest List of Mother's Day Gift Ideas for Under $15 http://www.facebook.com/group.php?gid=61364685955

Understand that we were in the middle of a recession, so this group was right on time.

I sent the invite to all of the people in my network. Some joined the group and some did not. People that were not in my network, but in the network of some of my friends, also joined the group. I created a list of 5 items. 2 were mine and one was a product that my sister was selling on her website. I encouraged people to add other items, but no one did. At the end of the day, I got the word out about my book to 76 people for **free**. It had been my experience that the success of a marketing strategy is hard to quantify. There are ways to track how marketing affected your web traffic and online sales, but this is not an exact science to determine your return on investment (ROI).

Facebook Fanpage Example:

A fanpage is similar to a profile page but it is for, as the name suggests, "fans" of a person, service, ideology, or product. It is a great tool to post information that is strictly related to your business, and also to separate people who might be interested in only your business and not you personally.

It is important to understand that these tools should not be your only vehicle for customer or client retention. We can use the Ice Cream Melts' Fanpage as an example. A large number of people identify with the concept of Ice Cream Melts and the types of projects that we develop. We created the Fanpage after launching the interactive version of our website. We put a link to our Fanpage on our website and people started to join.

I also sent out an internal Facebook message to my friends. Similar to the group example, some joined and some did not. Friends of my friends joined as well. There are people who are fans of Ice Cream Melts who are not in my personal network. This Fanpage enables me to post bulletins of our recent events, news articles, book releases, videos, website updates, etc. I can send out group messages or just post it to the profile of my Fanpage. Once again, this is a useful tool, but not our only tool.

Here is a link to the Ice Cream Melts Fanpage:
http://www.facebook.com/IceCreamMelts

Twitter (http://twitter.com/StrategicGen) – is an interesting concept that allows you to post short updates on your current status, similar to Facebook. The main differences are that you are limited to 140 characters and you can only post updates. Similar to other social networking platforms, you get a profile and the ability to customize it. You can also connect with friends known in Twitter as "Followers." Basically you can follow people or businesses and people can follow you. This means that you can see their updates and they can see yours.

Here are some things that I did with my Twitter page: (http://twitter.com/StrategicGen):

1. Followed a Twitter application that recommends to other people who share similar interests to follow me. Once your Twitter page is set up, the application can be accessed by going to: (http://www.mrtweet.net or http://twitter.com/MrTweet)

2. Connected my Twitter page to my website so visitors can choose to follow us by simply clicking the link. You can add your Twitter URL to your website or embed a widget that lists your latest Twitter posts.

3. Follow most people who follow me. Guess it is the rule of reciprocity. This seems to apply to people who are not incredibly famous. For instance, Ashton Kutcher (as of 06.11.2009) has 2,124,444 followers but is only following 167 people. Oprah has 1,405,047 followers, but is only following 14 people one of which is Ashton Kutcher.

4. Post things on Twitter pertaining to our business.

Here is a useful command from Twitter's Help Page:

@username + message — allows you to send a tweet to another person, and causes your tweet to save in that person's "replies" tab. "Tweets" mean messages in Twitter lingo.

Example: @StrategicGen I love that song too!

Visit the following for additional information on sending commands:

http://help.twitter.com/portal

There are plenty of other social networking websites. The moral of the story is that they are free tools. Use them as time permits to promote your business from within your network, and then expand outwards. Also, figure out ways to take advantage of the viral nature of these social networking tools. Technology is about innovation. There will be advances in social networking, and these tools discussed may be dwarfed, in time, by another tool. It is important to keep abreast of such technologies for the purpose of leveraging them for your business.

CHAPTER 7

GOOGLE WEBMASTER TOOLS AND ANALYTICS

G oogle is a great company with tons of free tools that can help your web initiative. The scope of this book will not go in depth about all of the tools offered by Google, but we will examine two tools that can get you started on the right path.

The two tools that we will cover are:

1. **Google WebMaster Tools** – provide you with detailed reports about your pages' visibility on Google.

2. **Google Analytics** – provides statistical usage data of your website. This includes how your site was found, how many unique visits, and how much time was spent on each page. With this information, you can improve your website's return on investment, increase conversions, and make more money on the web. This guide can help you familiarize yourself with the main features of Google Analytics.

More information on the Google WebMaster Tools and Analytics can be found on Google's Help Pages. Now it is time to really roll up your sleeves and get involved.

Google Webmaster Tools

1. Create a Google account by creating a Gmail email address. If you currently have one, then you can use that. Go to http://www.Google.com and click on Gmail.

2. Go to http://www.google.com/webmasters/start and sign in when prompted.

3. Click the "Add a site" button to get started.

4. You will be prompted to enter a URL:

    i.e. http://www.StrategicGeneration.com

5. Your website will need to be verified, meaning Google needs to make sure that you are the owner of the website. I recommend using the HTML file Verification method because it is easier. Google will give you a file name to create.

6. Create and Upload that file to your server. In order for Google to access the file, you must upload it into the /public_html folder of your site.

HINT: The easiest way to create the file is to make a copy of an existing html file and rename it to the filename provided by Google (i.e. googlea736577e68e585cb). No need to add the .html if you rename an existing html file

HINT: Upload files using the ftp program Filezilla that we discussed in *Chapter 4: Basic Goals of A Website*

7. Once the file has been uploaded correctly, click the Verify button.

8. The Next screen should state, "Congratulations! You're a verified owner of this site: 'http://www.Your_Site_Name. com.'"

If for some reason you are unable to verify your site, revisit the above steps.

A common issue is that your filename has a double html extension, i.e. googlea736577e68e585cb.html.html.
Simply remove the last .html and verify again.

HINT: You should be able to access Webmaster Tools by going to http://www.google.com and clicking on "Sign in" and then "My Account." Webmaster Tools should now be listed as one of your options.

Google has a wealth of Webmaster tools that give you the ability to understand how your site is being accessed by Google. This is, of course, important when people are trying to find you on the web. It will take another book to go in depth about Google's many tools. Now that your site has been verified, I urge you to submit a sitemap so that Google can crawl through your site and search for information queried by the millions of people who use the popular search engine. Google has help pages dedicated to walking you through creating and submitting a sitemap.

Don't get intimidated. Granted that the help pages will not be as easy to follow as the 8 steps above, it is important that you train yourself to understand such help pages. Once you are able to follow steps on technical help pages, more tools become available to you. There are tons of free resources on the internet that provide instructions on how to use them. Some may be more difficult than others, but at the end of the day they are **free**.

Google Analytics

1. Go to http://www.google.com/analytics/

2. Click the "Sign Up" button and enter your Gmail or Google account name and password

3. You will access the "Analytics: New Account Signup" page. Enter the appropriate information.

   Website URL: YOUR_DOMAIN_NAME

   Account Name: This will be the account name for this URL and Future URLs.

   Time zone country or territory: Self Explanatory

   Time Zone: Self Explanatory

4. Google will prompt you through a series of screens asking for your contact information and acceptance of their user agreement.

5. Google will then give you a code snippet to copy onto every webpage that you would like to track. This now gets us into development. If you are familiar with HTML and JavaScript then GO FOR IT!

   If not, have your web developer add the code snippet onto every page of your website. If your site was built using a content management system, as previously recommended, then there should be an automated process to get the Google Analytic code on all of your pages.

Instructions From Google:

Copy the following code block into every webpage you want to track immediately before the </body> tag. If your site has dynamic content, then you can use a common include or a template file.

Here is the main reason to have Google Analytics installed:

View in-depth monthly statistics on site visits, page views, bounce rate, average time on site, percentage of new visitors, location of visitors, content viewed, and because it is just **cool**. It feels good when I talk about www.IceCreamMelts.com and say that our average user spends 4:03 (min:sec) on our site, we had 524 visitors last month and 300 of them were new. People look at me like, wow, this guy knows a lot about his site.

Besides the random conversation at networking events, our analytics resulted in site updates and at one time, a complete overhaul of many pages. You can essentially track how visitors utilize your site. Version 2.0 of our site had videos on the homepage. We noticed that the majority of users will watch our videos and then leave our website. They would not drill down and look at the other things that we had to offer. We then switched and put video still pictures on the homepage. This then required the user to click on the image link and go to the video page to view the videos. Once they were on a secondary page they watched other videos. We would not have made this change if not for our Analytics.

Essentially, it is totally your call, but there is no way for you to know how people are using your site without some sort of tool that will help you analyze usage. Google Analytics is a good tool, and it is **free**.

Here are some additional **free** and useful Google tools:

Google Docs (http://docs.google.com) – This tool is great for collaborative document sharing. All members of your team can view documents and make changes online.

Google Voice Mail (http://voice.google.com) – This service allows you to receive all calls through a single number that will be redirected to your cell phone or land line. The service also includes call blocking/screening, voice mail transcripts, call conferencing, international calling and much more. There is currently a waiting list for new users to sign up for this service.

"SHARING information with OTHERS is the best way for useful INFORMATION to come back to YOU."

— *Nnamdi G. Osuagwu*

CHAPTER 8

FREE AND CHEAP RESOURCES TO ADVERTISE YOUR BUSINESS

Y ou have come a long way. Now it's time to generate traffic to your website. There are many books and online resources dedicated to marketing and advertising your business on the internet. We will cover a few resources and concepts to get you started.

The first type of advertising that you can do on your website is making sure that your content is informative and uses words that are industry standard terms in your market sector. It is important to be extremely descriptive in all of your service offerings and products. Some successful businesses have short descriptions of various services on their home page with a "Read More" link that the user can click to get more information.

Search Engines have functionality to crawl through content and find the most relevant information for the user doing the search. In the previous chapter, your website was added to

Google's webmaster tools along with a sitemap. This makes it easier for Google to find your website and scroll through your site's content when a user is requesting information on a particular topic.

There are also behind the scenes tags that are not visible on the webpage, but can contain information pertinent to your site. They are called meta tags and are inserted into the top portion of your webpage but can only be viewable in the actual html code of your webpage. Find the "view source" option of your browser to see the html code.

Here is some basic background on web browser functionality. One of the main jobs of the browser (i.e. Internet Explorer, Chrome, Firefox, etc.) is to translate html code and render the output. Some things in the code are not viewable when rendered, and some things are viewable.

Meta tags are not viewable, but can assist the search engine in locating websites that contain information that is being queried by a user.

The most commonly used meta tags are

1. **description** – short description of your site or page. Around 25 - 30 words are recommended.

2. **keywords** – main words that describe your site and page. Around 10 words are recommended.

Syntax

<meta name="description" content="description goes here">
<meta name="keywords" content="keywords go here">

It is a good practice to use the same keywords in both the meta tags and also within the content of your webpage. This practice will increase your chances of being found when someone searches for industry standard keywords within your business domain.

There are many pages that contain more information on meta tags. Here are a few:

How To Use HTML Meta Tags —
http://searchenginewatch.com/2167931

Meta element — http://en.wikipedia.org/wiki/Meta_tag

Free Resources to Advertise Your Business

Below is a list of some **free** online resources that you can use to advertise your business.

Craigslist.org (http://www.Craigslist.org) – This is a relatively free online message board that allows users to post services, products, events, etc. within their local vicinity. I mentioned relatively free, because they charge employers a fee in some cities for job postings. We use this service consistently for Sensible Developments, LLC to post our apartments as they become available. We even lucked out and purchased our capping machine on Craigslist. Previously, we were renting the expensive tool from Home Depot and finally lucked out and found a general contractor selling a slightly used one for a decent price.

Press Releases — Press releases or news releases are great tools for releasing information about a new product or service. I always thought this tool was reserved for major companies and that my information was not newsworthy. I learned differently after reading, *The New Rules of Marketing & PR* by David Meerman Scott. I now issue press releases for things pertaining to my business ventures. My most recent releases have been centered on new book projects. If your press release is well written and contains keywords specific to your industry, then it will be easier to find when someone is looking for information relating to your industry or service.

Jennifer Mattern did an excellent job on her website, http://nakedpr.com, of putting together a list of **free** press release services along with some useful advice. The information can be found by visiting the following links:

http://nakedpr.com/2007/07/16/effective-free-press-release-distribution-in-5-easy-steps

http://nakedpr.com/2007/07/29/big-list-of-free-press-release-distribution-sites

Cheap Resources to Advertise Your Business

Facebook Advertising (http://www.facebook.com/ advertising) —This is an inexpensive advertising tool. It allows you to target specific groups of people and advertise directly to them. You determine your daily spending budget and the amount of time you want your ad to run. You can either choose Cost-Per-Click (CPC) or Cost-Per-Thousand-Impression (CPM).

For my 2nd book, A Souvenir For My Mom, I ran a $2.00 a day ad with a .25 bid per couple of thousand impressions. One of my target groups included Obama supporters between the ages of 40-50 in the United States. I also ran one for users in London. It resulted in my site being trafficked internationally. I looked at my Google Analytics report and was able to the see the results of our Facebook ad campaign.

PRWeb - Press Releases (http://www.prweb.com) — You may pose the following question: Why pay, if other press release services offer it for **free**? Excellent question! Essentially, you will get more out of your press release with PRWeb's cheapest plan. It is all about your budget. My goal with this book is to give you **free** and inexpensive resources. My advice is to use as many **free** ones as you can and also inexpensive services, like PRWeb, depending on your marketing budget.

There are of course, plenty of other free and cheap resources online. The above is a good start. Also, all of my books have an online component where readers can interact;

this book is no different. In the *Let's Keep In Touch* section of this book, I will list a blog where I will be adding additional resources as I come across them.

Google Adwords (http://adwords.google.com) – You should have created a Google Account or Gmail email address for Google Analytics and Webmaster Tools. If you have not, you will need to do it for Adwords. This advertising method is similar to Facebook Ads, but instead Google and their many outlets (web and mobile) are part of your playground.

Here are some tips that can help you:

- An Ad Campaign must be set up initially. Think of it like the umbrella under which all of your Ads will fall.

- Choose the location(s) that work best for your campaign. You can choose to concentrate your campaign in a particular area (i.e. New York City) or go for a larger area (i.e. United States)

- Choose the appropriate bidding options. Cost-per-click (CPC), Cost-Per-Thousand Impressions (CPM), or Cost Per Acquisition.

Quick Definitions:

Cost-Per-Click (CPC) — You are bidding per click.

Cost-Per-Thousand Impressions (CPM) — You are bidding per 1,000 times your Ad is shown

Cost Per Acquisition — Google tracks which search queries produced higher conversions (actual clicks) and places higher bids on more favorable queries related to your campaign. Google is essentially storing and utilizing valuable data for your ad campaign. In order to use this form of bidding, your campaign must have received at least 30 conversions in the last 30 days, and you have to set up Conversion Tracking on targeted pages within your site.

To Learn More About Bidding Options:

https://adwords.google.com/support/aw/bin/answer.
py?answer=99484&ctx=tltp

- Once the Campaign is set up, then you must create an Ad Group for each ad that you will run. The Ad Group is the actual ad.

- Select your Bid Budget. Let's clearly understand how bidding works. Essentially your campaign is competing with other advertisers using similar key words. So when a keyword is entered by the user, the advertiser whose bid amount is higher will appear.

CHAPTER 9

VIRTUAL MAILBOXES

O ver the years, I have used my home address for business purposes. I also constantly moved and had current business mail sent to previous addresses. A virtual mailbox eliminates this issue and many more. A virtual mailbox is a service that acts as your business address. You can receive physical mail, which can be forwarded to your current address or scanned and emailed to you.

Here are some virtual mailbox services:

Earth Class Mail (http://www.earthclassmail.com)—
This company provides mailing addresses in Manhattan, Hollywood, Los Angeles, San Francisco, Seattle, and Portland. All addresses except for Portland incur additional fees on top of their standard monthly rates, which range from $19.95 to $59.95.

Their rates can be viewed at http://www.earthclassmail. com/pricing. Their service includes mail forwarding, scanning, storing, recycling, shredding and much more.

Forward Nevada—(http://www.forwardnevada.com/ MailForwardingServices.asp) — This service provides a Nevada Address and forwards your mail to you. The most cost effective approach is to pay for 12 months, which equates to $270 per year or $22.50 a month. You will also need to prepay a $50.00 postage deposit for mail forwarding postage fees.

Virtual Office— (http://www.regus.com/virtual) — In addition to a mail forwarding service, this company offers a telephone answering service and actual office space for meetings. Their prices range from $49 to $219 per month. They have locations all across the United States. Additional fees are incurred for major cities like New York and Los Angeles.

The above mentioned virtual mailbox services are just a few in this growing business. Here are some similar services to explore:

- The UPS Store (http://www.theupsstore.com/products/maiandpos. html)

- Private Box (http://www.privatebox.co.nz)

- Mail Link (http://www.maillinkplus.com)

- USA2Me (http://www.usa2me.com)

CHAPTER 10

CONCLUSION

This book covers merely the first steps in the long and fulfilling journey of business ownership. The most important aspect is to keep your spirits high and focus on giving the best product or service to your customers.

New businesses can benefit from starting off on the right foot with the tips discussed in this book, and existing businesses can integrate some of the tips into their current operations.

The last tip that I will talk about is something that was mentioned in *Chapter 2: Opening a Business Bank Account.* I referenced the usage of my Blackberry and Flash Drive.

It is important to always be accessible and create an image that your business is being manned at all times even if it is **not**. Below are additional tips on a couple of gadgets that can keep your business virtual and accessible:

Smart Phone Device (i.e. BlackBerry or iPhone) – There are plenty of smart phone devices on the market. Personally, I'm a Blackberry user. These devices allow you to send and receive emails from multiple email addresses. It creates an image of accessibility and availability. When a clients or customers send you emails, you can immediately get back to them. The image of accessibility and availability is easy to exhibit when you can receive and send emails from your cell phone device. At times, I even emailed important documents to myself before meetings. If the person whom I was meeting had a computer and printer, I could just forward the email to that person and ask that person to print it.

Flash Drives — Flash drives are mini storage devices that connect into the USB port of your computer. They can be easily carried in your pocket and are excellent backup devices. I sometimes carry a 4 gig flash drive on me, even when I don't have my laptop. It allows me to access my files from any available computer.

There are plenty of additional tools, but those two have kept me from killing defenseless trees and they also allow me to be constantly accessible and available.

Good Luck and I hope that the resources discussed in this book are beneficial to your business.

LET'S KEEP IN TOUCH

I'm interested in hearing about your
new or existing business.

Please comment on our Strategic Generation blog,
http://www.StrategicGeneration.com/blog.

Strategic Generation offers some of the technology services
discussed in this book.

Our website address is http://www.StrategicGeneration.
com.

Please send all inquiries to info@StrategicGeneration.com.

We can also be followed on Twitter :
http://twitter.com/StrategicGen or @StrategicGen

ACKNOWLEDGMENTS

An old adage, "A chain is only as strong as its weakest link," holds especially true in business. I would like to take this time and thank my different business partners who are integral to the companies that were mentioned throughout this book.

Ice Cream Melts, LLC & Global Boredom – *Nathaniel Thomas*

Sensible Developments, LLC – *David A. Shanks*

Strategic Generation – *Prabhat Sandheliya*

WhyGoOut, Inc. – *Michael J. Warthen*

I would also like to thank the following people
for their assistance with this project:

Sonia Canzater

Keah Buck

Darlene Gomez

Last but not least, I would like to thank the two people
that are my greatest source of motivation:

Grace Osuagwu (The Beginning)

Adanna Osuagwu (The Continuation)